Unicellular Organisms

L. Patricia Kite

Raintree

Chicago, Illinois

Editorial: Megan Cotugno, Andrew Farrow, and Abby Colich
Design: Philippa Jenkins
Illustrations: KJA-artists.com
Picture Research: Ruth Blair
Production: Alison Parsons

Originated by Modern Age
Printed and bound in China by Leo Paper Group

13 12 11 10 09
10 9 8 7 6 5 4 3 2 1

Library of Congress Cataloging-in-Publication Data
Kite, L. Patricia
 Unicellular organisms / Patricia L. Kite.
   p. cm. -- (Sci-hi: life science)
 Includes bibliographical references and index.
 ISBN 978-1-4109-3243-3 (hc) -- ISBN 978-1-4109-3258-7 (pb) 1. Unicellular organisms--Popular works. 2. Microorganisms--Popular works. I. Title.
  QR56.K578 2008
  579--dc22
                          2008026332

Acknowledgments
The author and publishers are grateful to the following for permission to reproduce copyright material: © Alamy/Paul Glendell p. **41**; Bettmann/Corbis p. **19**; © iStockphoto pp. **iii** (Contents, top), **13**, **31**; © iStockphoto/Joselito Briones p. **9**; © naturepl.com/Richard du Toit p. **33**; © Science Photo Library p. **28**, Andrew Syred p. **38**, CAMR/A.B. Dowsett p. **26**, CDC pp. **21**, **29**, Dr. Gary Gaugler pp. **6** (left), **6** (right), Dr. Kari Lounatmaa p. **24**, Dr. Linda Stannard, UCT p. **8**, Eye of Science p. **40**, Laguna Design p. **30** (left), Martin Dohrn pp. **23**, **32**, Mauro Fermariello pp. **15**, **20**, Mehau Kulyk p. **10-11**, Microfield Scientific Ltd. p. **36**, Michael Marten p. **39**, M.I. Walker p. **5**, Nancy Kedersha/UCLA p. **4**, Volker Steger, Peter Arnold Inc. pp. **iii** (Contents, bottom), **14**, Science Source pp. **6** (middle), **18**, Steve Gschmeissner pp. **17**, **30** (right); © Shutterstock background images and design features throughout.

Cover photographs reproduced with permission of © Science Photo Library/Mehau Kelyk **main**; © Science Photo Library/Dr. Linda Stannard, UCT **inset**.

The publishers would like to thank literacy consultant Nancy Harris and content consultant Dr. Michelle Raabe for their assistance in the preparation of this book.

Some words are shown in bold, **like this**. These words are explained in the glossary. You will find important information and definitions underlined and in bold, **like this**.

Contents

This early microscope first let humans see unicellular organisms.

Learn more about it on page 14.

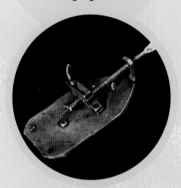

How do unicellular organisms help cows digest the grass they eat?

Find out on pages 13 and 31!

The Independent SINGLE Cell

Did you know that some life forms, such as **bacteria, are made of one single cell? These are called unicellular organisms. They can eat, breathe, and multiply on their own. The cell is the basic building block of life. <u>Every living thing around us is made of cells.</u>**

The First Life on Earth

Scientists believe that single-cell bacteria were probably the first form of life on Earth. Eventually, the cells clustered into very small groups. These organisms that are made up of more than one cell are called **multicellular organisms.** Humans are multicellular organisms. The cells of multicellular organisms depend on one another to stay alive.

This brain cell in the human body is a part of a multicellular organism.

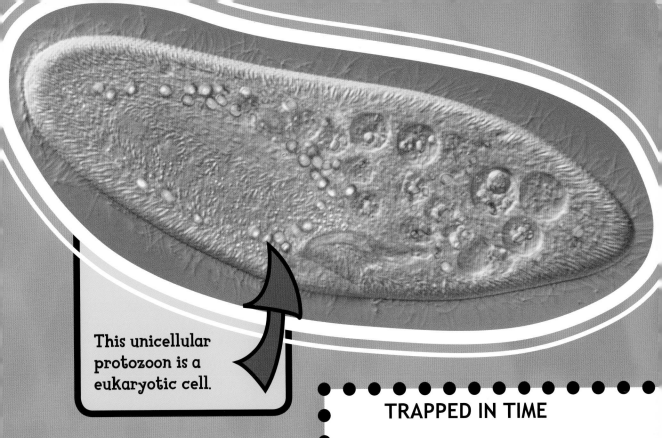

This unicellular protozoon is a eukaryotic cell.

Types of Cells

All cells are either **prokaryotic** or **eukaryotic.** The major difference between the two is that eukaryotic cells contain a **nucleus.** The nucleus holds the **genetic** material, or information, for the cell. Eukaryotic unicellular organisms include **fungi, algae, and protozoa**. All multicellular organisms are made up of eukaryotic cells.

Prokaryotic cells do not contain a nucleus. **<u>Bacteria are unicellular prokaryotic cells.</u>**

TRAPPED IN TIME

Amber is fossilized tree resin, or sap. Resin can preserve trapped creatures for millions of years. **Microbiologists** Raul J. Cano and Monica K. Borucki found a bee trapped in resin. They discovered that it was over 25 million years old. In 1995 they were able to bring cells from the bee back to life. The scientists claim these ancient cells' genetic material are similar to the genetic material of a modern strain of bacteria.

Testing showed that this bacteria survived without air and water. It also did not move or reproduce. Fascinating!

BACTERIA

There are at least 10,000 known **species**, or types, of **bacteria**. Bacteria live in the air, on land, and in water. They live on all types of plants and animals, including people, even though you cannot see them without a **microscope**.

Bacteria are **unicellular organisms**. Each bacterium is a single **cell** capable of all life functions, such as breathing, eating, and reproduction. Some move about using **flagella**, or little whip-like strands that act as propellers.

<u>**The three basic shapes of bacteria are round, spiral, and rod-shaped.**</u> In scientific studies, you will see these names in Latin. The bacteria with their singular and plural Latin names are below.

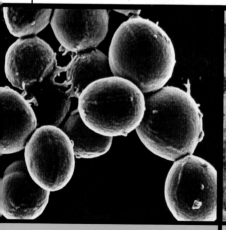

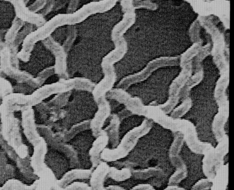

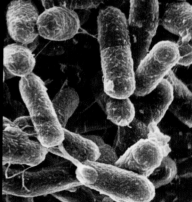

round bacteria

round = cocci/coccus

spiral bacteria

spiral = spirochetes and vibrios

rod bacteria

rod = bacilli/ bacillus

What's In a Bacteria Cell?

All bacteria are surrounded by a **cell membrane** and a **cell wall**. Inside is a jelly-like substance called **cytoplasm**. This contains everything in the cell. An important structure is the **DNA** (deoxyribonucleic acid). DNA is like an instruction manual for each cell; for example, it provides the number of **ribosomes**. These help the cell build the proteins that do many jobs.

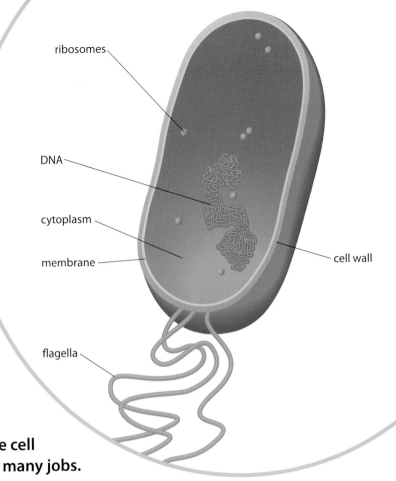

ribosomes

DNA

cytoplasm

membrane

cell wall

flagella

When a bacterium reaches full size, it begins multiplying. First, a copy of the DNA is made within the cell. Then, the cell divides into two identical cells.

Take a pencil. Make a tiny dot on a piece of paper. Can you guess how many bacteria will fit in that dot? (If you said between 500 and 1,000 bacteria, you are correct.)

QUESTION?

How can a simple scratch, a sip of dirty water, or a bite of rotten food cause you problems? Read on to find out.

Bad Bacteria and Good Bacteria

Bacteria cause a variety of illnesses and diseases in humans. These "bad" bacteria spread to humans many different ways—through food, insect bites, and contact with other humans. There are also "good" bacteria, including the bacteria that helps us digest food.

Food Poisoning

On a nice, warm day, you prepare a picnic lunch with an egg salad sandwich made with mayonnaise. When you get to the park, you decide to go for a walk. By the time you are ready to eat, your food has been sitting in the sun for several hours.

It is 27° Celsius (80° Fahrenheit) outside. Bacteria multiply most quickly between 10 and 54° C (50 and 130° F). Any **food poisoning** bacteria that were in the sandwich have now multiplied and may have released dangerous poisons. Eating the sandwich may cause you to suffer from food poisoning. Symptoms of this are vomiting and diarrhea. This is how our bodies react to bacteria and bacterial poisons that were eaten.

Why are many foods kept in the refrigerator? This is because bacteria don't reproduce well at cold temperatures. However, they do not completely disappear. Freezing is a better way to slow bacterial growth, although it cannot completely stop it.

Did You Know?

Salmonella bacteria can contaminate meat, poultry, dairy products and other food. The source of contamination is either the animal itself or the food handlers. If food is not properly cooked before eating, the bacteria cause fever, vomiting, and diarrhea.

Read below to learn more about protecting your food against poisoning bacteria!

Preventing Food Poisoning

Why do you boil some foods? Boiling kills most bacteria in food. As food cools off, bacteria can resume multiplying or enter from the air.

The most likely foods that are a source of food poisoning are meats, including chicken, and dairy products. Here are some ways that food poisoning can be prevented:

- Wash hands before preparing food.

- Do not sneeze or cough over food.

- Cook meat, chicken, and eggs well.

- Wash all fresh vegetables and fruit.

- Keep frozen foods in the freezer until you are ready to eat them.

The Black Death

During the 1300s, the bubonic plague, also known as the Black Death, swept through Asia, Africa, and Europe. Over 75 million people died, including entire villages. The cause was a bacterium called *Yersinia pestis,* which multiplied into trillions of tiny unicellular bacteria. The symptoms were high fever and large painful lumps, called buboes, in the armpits and on the neck and groin. Bleeding into the skin and other body parts created black patches on the skin.

During this time, sewage was thrown from windows and garbage was piled up on the street. Rats roamed freely, feeding on this garbage. These rats had fleas. Some rat fleas carried the deadly plague bacteria. When a flea bit a person, it took some of that person's blood. It also left behind some of the plague bacteria. These bacteria then multiplied. Within a couple of days, that person became ill.

Using the Plague?

Plague bacteria have been used as a biological weapon during wartime. Historical accounts from ancient China and Europe mention that infected animal and human bodies were thrown into an enemy's water supply. Plague victims were tossed by catapult into enemy cities.

Discovering the Real Cause

Before **microscopes** were invented, people thought plagues were caused by God's will, piled-up waste matter, unburied bodies, or the alignment of the stars.

In 1894 Alexandre Yesrin, a French-Swiss physician and **bacteriologist**, was a co-discoverer of the bacteria that caused the plague. He was the first scientist to track the transmission of the disease from rats to humans. He worked on an **antiserum** (a fluid that would kill the bacteria) for the disease. The disease is named *Yersinia pestis* in his honor. A Japanese physician and bacteriologist, Shibasaburo Kitasato, is also credited with discovering the bacteria.

The Black Death killed over 75 million people in the 1300s.

Good Bacteria?

Not all bacteria are harmful. **The human body requires many types of helpful bacteria in order to function.**

For example, the intestines, which move food from the stomach to the rectum, contain many helpful bacteria. The large intestine, which is warm and moist, contains about 400 to 500 bacteria species. There are more bacteria here than in any other part of the body. What do these bacteria do? As food moves through, bacteria break it down even further. **Helpful intestinal bacteria destroy many pathogens, or harmful substances, that may have entered your body.**

When you have a bowel movement, the feces contain food or other materials the body doesn't want or need at the time. Bacteria make up about 30 percent of feces. Unfortunately, some of the pathogens may not have been destroyed. That is why you must always wash your hands, with soap and warm water, after using the bathroom.

Did You Know?

In 1989 the *Exxon Valdez* oil tanker struck a reef off the coast of Alaska. This caused 11 million gallons of oil to spill into the Prince William Sound, which was home to a variety of wildlife, including sea otters and seals. Officials tried several different techniques to clean up the oil. One of those techniques was a **fertilizer** that encouraged a bacteria that would eat **hydrocarbons**, a chemical substance in **petroleum**. This procedure was successful in cleaning up some of the oil.

Trillions of bacteria live in the stomachs of cows, helping them digest grass.

Other Helpful Bacteria

Bacteria help keep the planet clean, too. Bacteria help break down dead animals and plants into tinier pieces. After a while, this decay becomes part of our Earth. While some bacteria cause food to go bad, others increase our food supply. Cattle need trillions of bacteria in their stomachs to digest the grass they eat. Bacteria also help change milk into cheese.

Other helpful bacteria are used in wastewater treatment plants. Toilet flushes and shower water both become wastewater. Scientists add helpful bacteria in special wastewater tanks to help clean up the water. After **disinfectant** is added, the cleaned water is used to spray off roads.

Discovering Bacteria

The year was 1648. Antoni van Leeuwenhoek had to earn a living. So, the Dutch teen became an apprentice (studying to learn a certain job) to a cloth merchant and learned to examine woven material. He used a magnifying glass to count the amount of thread in a cloth piece.

A Bacteria Detective

Leeuwenhoek began making larger magnifying glasses. He saw a magnifying glass set up on a small stand. He began creating his own simple **microscopes**. He used his microscopes to examine samples of lake water. "I saw therein, with great wonder, incredibly many very little animalcules, of divers sorts…." Some of the "animalcules" were green, clear, grey, or white, and round, spiral or rectangular.

Leeuwenhoek drew detailed pictures of his discoveries. They included **bacteria**, **protozoa**, **fungi**, and **algae**. He wrote to famous scientists describing exactly what he saw, but few believed him. To prove the single-celled organisms were alive, he demonstrated killing them with vinegar. In 1680 people finally accepted the existence of these **unicellular organisms**.

Antoni van Leeuwenhoek viewed unicellular organisms after developing his microscope.

LEEUWENHOEK'S MICROSCOPES

Leeuwenhoek's early microscopes first magnified three times what could be seen with the naked eye. Later ones could magnify up to about 275 times. Before his death, he had created 400 different types of microscopes.

When people could see **microorganisms**, they began understanding that bad deeds and magic were not the causes of disease. This began a new wave of research. Scientists began looking for cures to destroy the harmful microorganisms seen through the microscope.

Did You Know?

Electron microscopes use a beam of electrons (negatively charged particles) to magnify organisms or objects. Modern electron microscopes can magnify up to 2 million times. This scientist is looking through an electron microscope to study unicellular organisms.

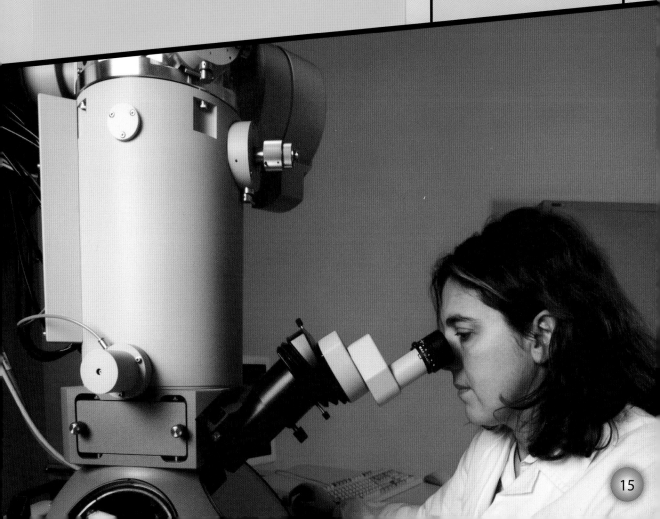

Fighting the Bad Bacteria

We come in contact with many harmful **microorganisms** every day. Some enter our bodies and make us sick. Our bodies usually fight off these illnesses, but that does not always work. In these cases, we turn to other things to make us better.

Body Defenses

The human body has many natural defenses against illness. When disease-causing microorganisms or "germs" enter a body, special germ-fighting agents rush to the area. Some of these agents are protein substances called **antibodies**. They may be present due to vaccination or a prior infection. The antibodies surround and attach to the invading germs. This initially keeps the germ from spreading. Then, **white blood cells** enter the area.

<u>**White blood cells are responsible for the body's defense system.**</u> They fight infections and are protection against many harmful invaders such as **bacteria** and **viruses**. There are about ½ million white blood cells in every drop of human blood.

Sometimes, microorganisms multiply too fast for the body to organize a good defense system. Other times, the defense system just fails to work, even with the help of modern medicines.

Disease-causing bacteria can infect a person in two ways. Some attack and destroy healthy body cells directly. Others produce harmful chemicals called toxins. These toxins act like poisons to destroy or damage healthy cells.

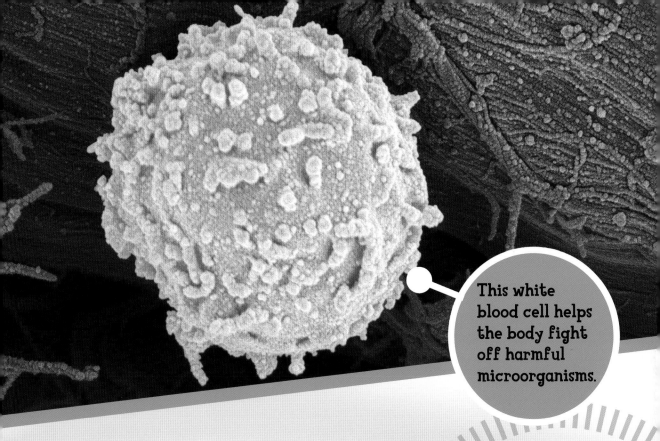

This white blood cell helps the body fight off harmful microorganisms.

Getting Infected

There are many ways harmful microorganisms can enter your body, making you sick. These include:

- Puncture wounds, such as a scrape or scratch on your skin

- Not washing your hands before eating

- Drinking contaminated water

Harmful microorganisms can enter the lungs, bloodstream, and mouth. From there, they travel throughout the body. Eventually, your body begins showing symptoms of being infected, such as a sore throat or nausea.

Alexander Fleming

Alexander Fleming, born in Scotland in 1881, worked as a doctor during World War I. He watched many soldiers die from infected wounds. When the war ended, Fleming became a **bacteriology** professor. He wanted to find some way to stop those with wound infections from dying.

Fleming's laboratory was rather cluttered, and sometimes he forgot to clean up. One day in 1928, he forgot to clean some of his **Staphylococcus** bacteria experimental dishes before he went on vacation. When he returned, some of the dishes had blue mold on them. Fleming threw them in a **disinfectant** container. He soon forgot about them.

Later, Fleming took out some of the dishes he had tossed in the disinfectant container. He picked the ones on top. They hadn't landed in the cleansing liquid. He noticed that the area near the mold didn't have any Staphylococcus bacteria.

Fleming studied the blue mold for four years. It seemed to be able to kill off harmful bacteria. He called it penicillin. Fleming wrote about penicillin, but the medical community paid little attention to his discovery.

Fleming kept trying to isolate the **antibiotic** penicillin from the mold. He wanted to produce a lot of it in a usable form. He didn't succeed, but other scientists were inspired to continue the research.

Gladys Hobby

From 1934 to 1943, **microbiologist** Gladys Hobby worked with a Columbia University Medical School team in New York City. They were trying to develop penicillin for several infectious diseases. She brewed the first batch of penicillin tested on humans. Hobby was also with the first research group to treat a patient with penicillin injections. She later developed Terramycin, an antibiotic more effective than penicillin against some diseases.

Over the years, penicillin has been successfully used against almost all bacteria harmful to humans. Diseases it has treated include **anthrax**, **tetanus**, and **scarlet fever**.

Super Bugs

Many new antibiotics have been discovered since penicillin. Doctors prescribe antibiotics to people when they are sick. It is important to take all the antibiotic, or all the bacteria may not be killed. Antibiotics are also given to farm animals, such as cattle, to keep them healthy. These antibiotics stay in the animals when they become food.

Staph Bacteria

The excessive use of antibiotics can cause bacteria to mutate, or change, becoming resistant (not dying) to them. A common bacteria today is called MRSA (methicillin-resistant Staphylococcus aureus). These Staph (short for Staphylococcus) bacteria are resistant to almost all current antibiotics. Estimates are that MRSA infection complications kill 18,000 people a year in the United States.

This researcher is studying Staphylococcus bacteria.

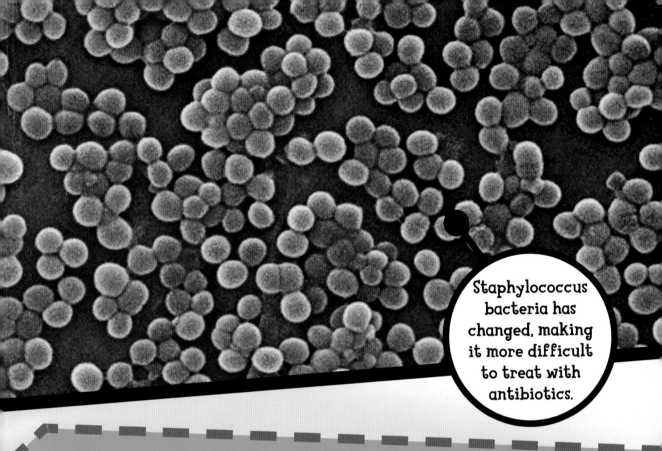

Staphylococcus bacteria has changed, making it more difficult to treat with antibiotics.

MRSA

In the recent past, MRSA infections were limited to patients undergoing surgery in hospitals. However, recently the resistant bacteria have changed even more. Now there is CA-MRSA, (community associated-MRSA). These bacteria cause severe infections in otherwise healthy people. Somehow, the Staph bacteria have learned to attack the body's protective cells.

Scientists are working hard to find a way to stop MRSA and CA-MRSA. In the meantime, protect yourself. Wash your hands with soap and warm water following restroom activities. You should report any sores to the school nurse or a physician.

All open wounds should be completely covered until healed. This is especially important if you are involved in physical contact activities. There has been a marked increase in MRSA among school-age children, especially student athletes. Student athletes should not share towels, clothing, or other personal items.

Before Antibiotics

Before modern medicine, people used many methods to cure illnesses. Physicians did not know the cause of a disease, so they treated the symptoms. Desperate people tried anything to stop the illness.

For a cold, goose grease and turpentine were mixed together and rubbed into a patient's chest. Sufferers of a sore throat wrapped a dirty sock around the neck.

Bloodletting, letting blood out of the body by cutting the skin, was a very common remedy until the late 1800s. One way was applying leeches onto a person's skin over the veins.

Bloodletting often made the patient sicker. Since blood contains the cells that fight disease, losing blood is actually very harmful. Also, physicians did not wash equipment between patients. Sometimes, they used the same medical equipment on people and animals without ever washing it. This caused people to develop further infection.

Did You Know?

For many years, doctors turned away from the practice of using leeches. However, recently, doctors have found that leeches can be helpful during certain procedures. Leeches' saliva helps increase blood flow to damaged areas of the body.

Leeches are also helpful in re-attaching body parts. Physicians have also developed mechanical leeches that have the same benefits without having to put an actual leech on someone's body.

The Health Detectives

One day you feel fine but the next you come down with a fever, nausea and headache. Do you wonder what exactly is causing this illness and how you got it? **Epidemiologists** are the people who work to find the answer.

Epidemiology is the science concerned with the study and control of epidemic diseases.

An epidemic is the appearance of an infectious disease or condition that attacks many people at the same time in the same geographical area.

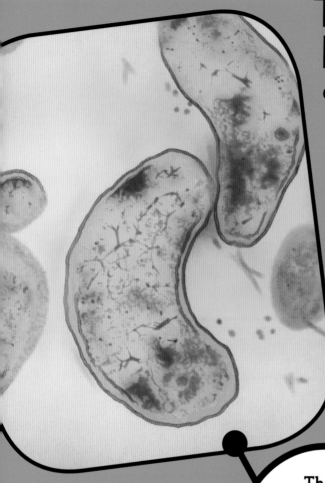

These Campylobacter bacteria are a common cause of illness in humans.

Campylobacter

Sarah O'Brian is Professor of Health Sciences and Epidemiology at the Manchester School of Medicine in England. For 10 years, she studied Campylobacter, a common illness-causing bacteria.

C. jejuni, one form of Campylobacter bacteria, causes fever, stomach pain, and headaches. It is the most common bacterial cause of stomach problems in the world. Where does it start? Its main cause is eating under-cooked chicken or drinking contaminated milk or water.

The Manchester study found that Campylobacter infections increased in late spring. Epidemiologists compared weekly temperatures with the number of reported Campylobacter infections. They found that for every one-degree rise in temperature there was a 5 percent increase in infection. This epidemiological study provided evidence that temperature influenced the number of Campylobacter infections.

QUESTION?

Have you ever been sick and then tried to trace back to where you caught the illness?

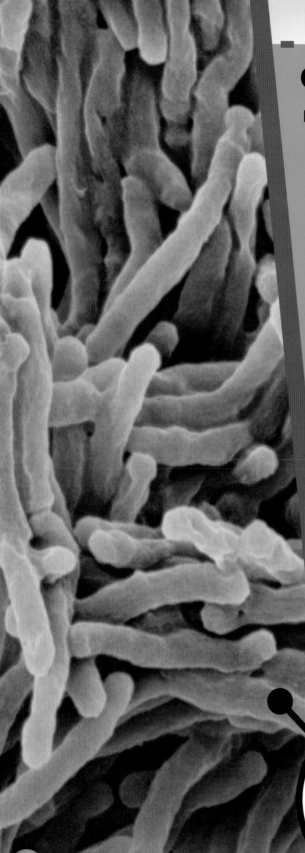

Cholera

In the mid-1850s, an outbreak of **cholera** began killing people in London, England. Cholera affects the intestines and causes diarrhea and dehydration. Most people believed breathing dirty air caused the disease. Dr. John Snow disagreed. He insisted the disease was caused by contaminated water. At the time, sewage was emptied into the Thames River. This was the source of water for local residents.

Snow made a map of the city. He marked each cholera death on the map. He then determined where each cholera victim got drinking water. Most deaths occurred in an area serviced by a particular water company. Snow later traced an outbreak of cholera to a public water pump. The water pump was eventually removed. It took a while for people to accept his ideas, but Snow's findings helped save many lives.

These tuberculosis bacteria can spread through the body and are fatal if not treated.

Tuberculosis

Tuberculosis is caused by a bacterium called *Mycobacterium tuberculosis*. (See the photo of the bacteria on the previous page.) Symptoms include coughing up blood, fever, and chest pains. Tuberculosis is spread when a person breathes in airborne droplets that are ejected when an infected person coughs.

Tuberculosis was a major cause of death in the 1800s. By the middle of the 20th century, antibiotics were used to treat the illness. It became less common. But in the mid-1980s, the number of cases began to increase once again. The bacteria had become resistant to the medicine once used to treat it.

Recently, tuberculosis cases have more than doubled in large cities. Epidemiologists wanted to know how tuberculosis was being spread, so they conducted a study. For this study, a group of tuberculosis patients reported to the San Francisco Department of Public Health for a one-year period.

The study found that the disease occurred more often in males, people with AIDS, and people of certain ethnicities. It also concluded that patients who do not receive treatment are more likely to spread the illness than those who receive proper treatment.

VIRUSES

Common colds and the flu are caused by **viruses**, not bacteria. Bacteria are **prokaryotic** cells that can reproduce on their own. Viruses are not cells, and they cannot reproduce on their own. They are protein capsules that contain **DNA or RNA**. Viruses need a **host cell** to reproduce.

How Do Viruses Live?

A virus is a parasite, because it needs a host cell to reproduce. Each type of virus can only infect specific host and host cells. First, a virus attaches to a host cell and injects its DNA or RNA. The virus then takes over and directs the host cell to make new viruses until it eventually dies. The new viruses invade other cells of the same host, or they leave to find a new host.

Viruses are much tinier than bacteria. They can only be seen through the most powerful microscopes. Yet they can cause severe diseases, including **AIDS, yellow fever, chicken pox,** and **hepatitis.**

This HIV (human immunodeficiency virus) particle causes AIDS in humans.

How Do Viruses Spread?

Viruses need to travel to new hosts. A cough or sneeze shoots thousands of viruses into the air. Viruses are also spread through infected blood, contaminated food and water, and by people not washing their hands after using the bathroom.

The body fights viral infections with antibodies and immune cells that hunt and kill infected host cells. The body's immune system remembers most invaders and can clear a second infection very quickly. The flu virus is tricky because every year it changes its appearance. When the body is attacked by flu, its protective cells have to start over again.

Vaccinations

Vaccinations trigger the body's antibody protection for many bacterial and viral diseases. Some vaccines are made from germs that have been scientifically made much weaker in a laboratory. Vaccines allow your body to learn about a disease so that if it ever sees it again, it is ready to fight. Some viral diseases, such as **smallpox**, have now been eliminated due to vaccination.

Scientists are working very hard to find vaccines against more diseases. Until they succeed, you can help protect yourself against viruses in the same ways as against bacteria. This includes:

- Covering your mouth when sneezing and coughing.
- Staying home when sick.
- Not drinking food or water that might be contaminated.

This yellow fever virus can be prevented in humans by vaccination.

PROTOZOA

Protozoa are unicellular organisms. Most live in moist places, including soil and inside animals. Protozoa have many different types and shapes.

Protozoa move in different ways. Some move by using whip-like **flagella**. Other protozoa have body **cilia**. These hair-like extensions work somewhat like oars. Some protozoa move with **pseudopods**, or "false feet." There are also protozoa that just drift. Most protozoa multiply by **binary fission**, or separating into two equal parts.

Some protozoa feed on animal or plant material, while others hunt for food. Many protozoa absorb food through their **cell membrane**. Others get food through a mouth-like structure. If surroundings are unsatisfactory, some protozoa can create a barrier around themselves. In this **cyst** form, they rest until living conditions get better.

These protozoa move with their cilia.

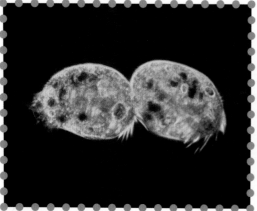

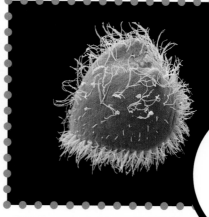

This protozoon is found living in fresh water.

Protozoa live in these termites to help them digest the wood they eat.

Good and Bad Protozoa

Some protozoa are beneficial, such as those that live in the intestines of cows and sheep. These protozoa help break down the tough cellulose, or material, in grass. This helps the animals digest their food. Without these protozoa, the animals would starve.

Protozoa living within termites help them digest wood. In the wild, termites help break down fallen trees into soil particles. Other protozoa help break down old logs, fallen leaves, and dead animals into earth compost.

Protozoa can also cause disease. There are protozoa that cause **sleeping sickness** and **malaria**. Read on to find out more.

Sleeping Sickness

Protozoa cause the deadly disease sleeping sickness. Sleeping sickness affects over 60 million people in 36 countries. This protozoon is carried by the blood-sucking tsetse fly, which lives in sub-Saharan Africa. If treatment is prompt, chances of survival are good. However, almost all untreated patients die.

The cycle begins when the tsetse fly bites an animal that is carrying the protozoon called trypanosome. Many cattle and other wild animals may be carriers. When the tsetse fly bites, it also gives off the protozoa. The protozoa multiply in the bloodstream and body tissues.

The symptoms of the disease are fever, large **lymph glands,** headache, and joint pain. The victim sleeps during the day and is awake all night. Eventually, the brain becomes so damaged that the patient slips into a coma and dies.

The tsetse fly carries the protozoon that causes sleeping sickness.

Fake cows like this one help reduce the number of tsetse flies.

Fake Cows

Many efforts are being made to eliminate the tsetse fly that causes sleeping sickness. They are also working to eliminate **nagana**, the form of sleeping sickness that affects animals. An international group of researchers designed an artificial cow. This cow doesn't look like a real cow, but it smells like one, attracting tsetse flies. It also contains **insecticides**, which eventually kill the tsetse flies that land on it.

After 60,000 artificial cows were introduced, the number of cattle with nagana disease dropped to almost zero. The number of tsetse flies dropped, too, so that fewer insecticides were needed.

THE INVESTIGATOR:
Dr. Louise Pearce

Massachusetts native Louise Pearce received her medical degree from Johns Hopkins Medical School in Baltimore, Maryland, in 1912.

Dr. Pearce particularly enjoyed field research. She and a team of scientists developed a medication against African sleeping sickness. But, it needed testing on people. When another severe epidemic of sleeping sickness occurred in the Belgian Congo in the early 1900s, Pearce volunteered to go alone. This area is now called the Democratic Republic of the Congo.

Dr. Pearce tested the medication on 70 patients. The sleeping sickness protozoa disappeared from their bloodstream within a few weeks. Later, the medicine cured more than 100,000 Africans.

MALARIA

Malaria kills over one million people each year. It is one of the world's deadliest diseases. While most of the deaths occur in Africa, the disease is now spreading to areas once free of malaria. With increasing travel to undeveloped countries, many people now have contact with the *Anopheles* mosquito. This mosquito carries the protozoon, called plasmodium, that causes malaria. Travelers can become infected after being bitten by the mosquito.

Malaria victims have cycles of extreme chills and then burning fevers. The earliest description of malaria came from ancient China, around 2700 BCE. Many attempts were made to find a medicine that would help. Physicians tried **bloodletting**. People once believed malaria came from breathing stinky swamp air.

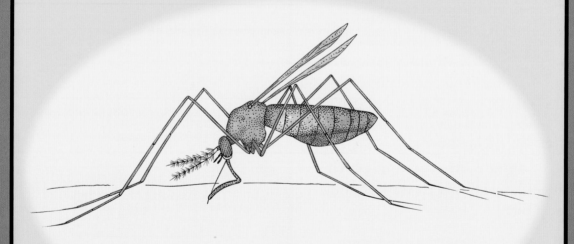

MALARIA DETECTIVES

Alphonse Laveran

Alphonse Laveran came from a military family. Like his father, he became a military physician. In 1878 he was sent to Algeria. He spent a lot of time looking through a **microscope**, trying to find the cause of malaria. He examined blood samples from living and dead patients. In 1880 he was the first to see the protozoa that caused malaria through a microscope.

Ronald Ross

Ronald Ross was born in India in 1857. He graduated from medical school and began world travel. Tropical diseases interested him, especially malaria. One day he dissected an *Anopheles* mosquito. In the 1890s, Dr. Ross was able to prove that the malarial protozoa parasite could transfer through mosquito saliva into the bloodstream.

Unfortunately, many mosquitoes have developed resistance to insecticides and treatment drugs. The good news is that scientists are continuing to work toward prevention and treatment.

PREVENTING MALARIA

What is done to prevent malaria?

- Closing windows and doors to prevent mosquito entry.
- Keeping outdoor water containers empty (since mosquitoes lay their eggs in water).
- Using mosquito-eating fish in fishponds.
- Killing pond mosquito **larvae** (young mosquitoes) with **larvicides**.
- Wearing protective clothing.
- Getting preventative medicine from a health clinic.

YEASTS

Yeasts are **microscopic single-celled fungi. The fungus kingdom includes organisms such as molds, mildew, and mushrooms.** There are hundreds of different yeast species. Each yeast cell eats, breathes oxygen, and acts on its own.

How Does Yeast Grow?

Yeast cells multiply by a method called budding. A smaller yeast cell, or bud, grows from the side of a parent cell. As the bud grows, it remains attached to the parent for about an hour. Then, it separates and begins its own budding process.

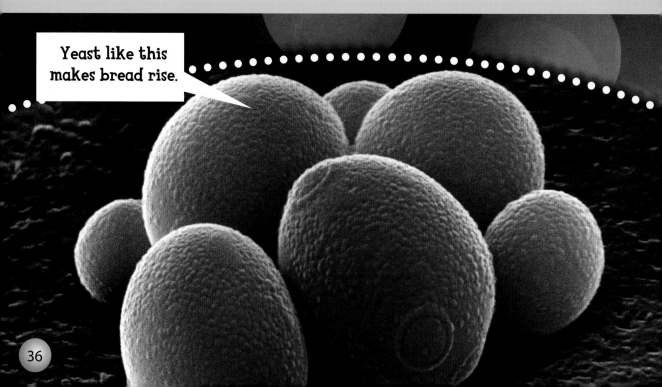

Yeast like this makes bread rise.

Good and Bad Yeast

Some yeasts are harmful to humans. These yeasts are **parasites**. An example of a parasite yeast is *Candida albicans*. This is the yeast that causes diseases in humans.

Other yeasts are helpful to humans. Yeast used in bread dough comes to life when mixed with warm water. It then eats the sugar in the dough, causing it to release air. This creates bubbles in the dough, which causes it to rise. The yeast die in a few hours.

What happens if yeast isn't added when making bread? Mexican tortillas, Jewish matzo, Indian chapatis, and Southwest Hopi Native American piki bread are all made without yeast.

TRY THIS!

If yeast cells are alive, what conditions do they prefer? Try this experiment to find out. Yeast can be bought from the grocery store.

Dissolve $\frac{1}{4}$ teaspoon of yeast in a little lukewarm water with a pinch of sugar added. If, within five minutes, the mixture foams lightly, the yeast cells are alive.

Does it make a difference what type of sugar you use? Experiment with brown and powdered sugar.

Now, using the same type yeast, dissolve $\frac{1}{4}$ teaspoon in cold water. What happens? What happens in hot water?

With an adult, try baking bread with, and without, yeast. Don't forget both clean hands and a clean working area.

ALGAE

Many algae are **unicellular organisms** that live in oceans, freshwater lakes, ponds, and rivers. Species of golden, green, and red algae are unicellular. These algae, such as seaweed, may look like plants. But these algae are not plants. They don't have roots, stems, or leaves.

Some single-celled algae use the Sun's energy to help create their food supply. Others eat whatever suitable food lives near them. Some stick out a feeding tube to suck in food. **Algae multiply by binary fission, where a cell separates into two equal parts.** Unicellular algae may move using whip-like **flagella**. Others may drift or use **pseudopods**, or false feet.

These algae organisms use flagella to move.

An algae bloom is harming life in this pond.

You may have seen a green, red, or brown layer forming on ponds or other water bodies during the summer. This "algae bloom" gets thicker as the water dries up. The thick layer blocks light below. Water plants die, and the dead organic matter becomes food for some **bacteria**. The bacteria multiply, using up the water's oxygen. With less oxygen, the fish die.

What makes algae reproduce to this extent? Scientists debate this. Many think pollution, such as fertilizer runoff, sewage water, and garbage, is the cause.

TRY THIS!

Under what conditions do algae grow the fastest? Try the following to find out. Take several clean baby food jars. Fill each ⅔ full of tap water. Place one in a dark place, one in a semi-light place, and one in full sun. Put in a tablespoon of fish tank water. Do not cover the jars. Observe the water each day through a microscope. Watch for algae. You can also try this activity using distilled water or purified drinking water. Does this change the results?

Slime Molds

There are many different kinds of molds. But slime mold is not like the mold you see growing on old bread. Some forms of slime mold are **unicellular organisms** that eat decaying wood, leaves, or other organic matter that you would find outside. Some scientists think they are **fungi** or a form of **protozoa**.

Slime molds reproduce through this fruiting body.

The Life of a Slime Mold

Slime molds live in cool, moist places on the rotting material that they are consuming. Slime molds can be yellow, orange, red, whitish, or clear. Some look like slime, others like fuzz. As the slime mold moves, it eats dead organic matter.

<u>**Plasmodial slime molds are a single cell.**</u> In times of drought, the plasmodial slime mold groups into bunches. Tiny egg shapes show up on tiny stalks. These are called fruiting bodies. Fruiting bodies can be orange, red, brown or purple. The fruiting bodies give off **spores**. Spores are their way of reproducing. Spores have a thick outside wall that lets them survive unpleasant environmental conditions. They may stay around rotting logs or move with the wind. If they land in more damp plant areas, new slime molds grow from them.

This slime mold is slowly moving across a log.

Classifying Unicellular Organisms

Scientists like to sort things into categories, or groups. **Living things are usually divided into five kingdoms.** Each kingdom is divided into small groups. These smaller groups are then divided until each group contains only one **species**, or type, of organism. However, depending on the scientist, there may be even smaller groups. Scientists don't always agree with each other on the groupings! **The five main kingdoms are Fungus, Animal, Protist, Monera, and Plant.** The three below have been discussed in this book.

Three of the Five Main Kingdoms		
Fungus	Protist	Monera
• Yeasts	• Protozoa • Euglena • Some algae • Slime molds and water molds (some scientists classify these as fungi)	• Bacteria • Archaebacteria (live in extreme environments)

Unicellular Organisms Review

◆ **Bacteria** are **unicellular organisms**. This means they are made from a single cell.

◆ All cells are either **eukaryotic** cells or **prokaryotic** cells. Eukaryotic cells contain a **nucleus**; prokaryotic cells do not.

◆ **White blood cells** protect our body against harmful bacteria and **viruses**.

◆ Unlike bacteria, viruses are not alive. The come to life by invading a living cell. Viruses cannot be treated by **antibiotics**, as illnesses caused by bacteria can.

◆ **Protozoa** are unicellular organisms. They move around using **flagella**, **cilia**, or **pseudopods**.

◆ Most protozoa multiply by using **binary fission**. This means they separate into two equal parts.

◆ Some protozoa cause disease such as **sleeping sickness** and **malaria**.

◆ **Yeasts** are single-celled **fungus**. Some yeast is used to make bread dough rise.

◆ Certain types of slime molds are unicellular organisms. Slime molds eat decaying leaves and other organic matter.

◆ Many **algae** are unicellular organisms. They multiply by binary fission.

◆ There are five main **kingdoms**: Fungus, Animal, Protist, Monera, and Plant.

Glossary

AIDS (Acquired Immunodeficiency Syndrome) A virus-caused disease that attacks a person's immune system

Algae Nonflowering, single-celled plants, usually living in water

Anthrax A bacteria-caused disease marked by itchy bumps on the skin

Antibiotic Substance that kills or slows the growth of bacteria

Antibody A chemical made by the white blood cells to fight off the invasion of harmful microorganisms in the body

Antiserum A fluid containing antibodies

Bacteria Unicellular organisms that live in all conditions. There are over 10,000 known species of bacteria.

Bacteriologist A person who studies bacteria

Bacteriology The study of bacteria

Binary fission Reproduction by separating into two equal parts

Bloodletting Cutting skin or a vein to let blood out of the body

Budding A form of reproduction in which a cell copies itself and produces an outgrowth before splitting in two

Cell Basic unit of life

Cell membrane Continuous layer of fat or protein that encloses a cell

Cell wall A layer surrounding a cell outside of the cell membrane that provides structural support

Chicken pox A virus-caused illness marked by a fever and spots on the skin

Cholera A bacteria-caused illness, affecting the intestines and marked by diarrhea and dehydration

Cilia Hair-like extensions that some protozoa use to move

Cyst Protective structure used for resting purposes by some microorganisms

Cytoplasm Jelly-like substance that fills a cell

Disinfectant A chemical solution intended to destroy harmful microorganisms

DNA (deoxyribonucleic acid) The part of a cell that holds the genetic information about the organism

Electron microscope A microscope that uses a beam of electrons (negatively charged particles) to magnify organisms or objects too difficult to see or too small to see with the naked eye

Epidemic The appearance of an infectious disease or condition that attacks many people at the same time in the same geographical region

Epidemiologist A person who studies epidemic diseases

Epidemiology The study of epidemic diseases

Eukaryotic Organism made up of a cell or cells that contain a nucleus

Fertilizer A substance used to improve soil for the growth of plants

Flagella Whip-like structure projecting from some unicellular organisms, used for movement

Food poisoning Illness caused by eating bacteria-infected food. Symptoms include nausea, vomiting and diarrhea.

Fungi Any group of spore-producing organisms

Genetic Having to do with all the information about and make up of an organism

Hepatitis A virus-caused disease marked by inflammation of the liver

Host cell Cell in which a virus makes copies of itself

Hydrocarbon An organic compound of hydrogen and carbon used in petroleum

Insecticide A substance used to kill insects

Kingdom The most general level in the classification of living things

Larvae An insect in its youngest stage of life

Larvicide A substance used to kill larvae

Lymph gland One of many small organs throughout the body that are a part of a system that produces chemicals the body needs to function

Malaria A protozoa-caused disease transmitted to humans by mosquitoes. Symptoms include cycles of chills and fevers.

Microbiologist A person who studies microscopic life

Microorganism Any microscopic living thing, such as bacteria or protozoa

Microscope An instrument used to magnify organisms or particles difficult to see or too small to see with the naked eye

Microscopic Something too small to be seen without the use of a microscope

Multicellular organism An organism made up of more than one cell in which the cells depend upon each other to survive

Nagana A protozoa-caused illness in animals passed on by the tsetse fly, similar to sleeping sickness in humans

Nucleus Central part of eukaryotic cells that contains the DNA

Parasite An organism that lives on another, depending on it for food and not giving anything in return

Pathogen A harmful microorganism or substance

Petroleum A naturally occurring liquid found in the earth used to make fuels

Plasmodial slime mold Unicellular organisms that store energy, live on land, lack cell walls, and eat bacteria, yeast, fungi and decaying organic matter

Prokaryotic Unicellular organisms that do not contain a nucleus, such as bacteria

Protozoa Single-celled animal organisms

Pseudopods False feet on some unicellular organisms used to move

Ribosomes Part of a cell that helps make protein

RNA (ribonucleic acid) A molecule found in the nucleus that is involved in producing proteins

Salmonella A type of bacteria that causes diarrhea; spread mostly by contaminated food

Scarlet fever A bacteria-caused disease marked by a red rash on the body

Sleeping sickness A protozoa-caused illness transmitted to humans by the tsetse fly. Symptoms include fever, large lymph glands, headache, joint pain. Victims sleep all day and are awake at night.

Smallpox A virus-caused illness marked by aches, fever, and spots on the skin that form pockmarks

Species Level of classification that contains similar organisms

Spore The reproductive part of certain organisms

Staphylococcus A bacteria of which some strains cause infections in humans

Tetanus A bacteria-caused disease marked by muscle spasms and lockjaw

Tuberculosis A bacteria-caused disease, marked by a fever, cough, difficulty breathing, and lumps inside the lungs

Unicellular organism A single-celled organism that can eat, breathe, and reproduce on its own

Virus A parasite that attacks a host cell and reproduces

White blood cell A cell found in the blood that helps the body fight off infection

Yeast A group of unicellular fungi

Yellow fever A disease usually found in Africa and South America that is caused by a mosquito. Symptoms include fever, chills, headache, muscle aches, nausea and weakness.

Further Information

Books to read

Bjorklund, Ruth. ***Health Alert: Food-Borne Illnesses***. Tarrytown, NY: Marshall Cavendish, 2006.

Claybourne, Anna. ***World's Worst Germs: Microorganisms and Disease***. Chicago: Raintree, 2006.

Claybourne, Anna. ***Microlife: From Amoebas to Viruses***. Chicago: Heinemann Library, 2004.

Nye, Bill. ***Bill Nye the Science Guy's Great Big Book of Tiny Germs***. New York: Hyperion, 2005.

Snedden, Robert. ***A World of Microorganisms***. Chicago: Heinemann Library, 2007.

Stille, Darlene R. ***Classifying Living Things***. Strongsville, OH: Gareth Stevens, 2008.

Websites

ThinkQuest's Microorganism Library
http://library.thinkquest.org/26644/us/index.htm
Lots of information about bacteria, fungi, protozoa, viruses, and the diseases that they cause.

Center for Disease Control and Prevention's Body and Mind Website
http://www.bam.gov/sub_diseases/index.html
Learn more about diseases and the people who are trying to stop them.

Microbe World
http://www.microbeworld.org/
Meet scientists, watch videos, and learn all sorts of things about unicellular organisms.

Digital Learning Center for Microbial Ecology
http://commtechlab.msu.edu/sites/dlc-me/
Learn about different microbes, microbes in the news and visit the microbe zoo.

Index